MW01232942

The KidHaven Science Library

Space Travel

by Don Nardo

KIDHAVEN
PRESS™

THOMSON

GALE

San Diego • Detroit • New York • San Francisco • Cleveland
New Haven, Conn. • Waterville, Maine • London • Munich

THOMSON

GALE

™

Picture Credits

Cover: © Victor Habbick Visions/Science Photo Library
© Associated Press/NASA, 11
© Bettmann/CORBIS, 17
© CORBIS, 8
Gazelle Technologies, Inc., 5, 15
Rick Guidice/NASA, 40
© R. Harris/Photo Researchers, 19
Chris Jouan, 9, 24, 34-35
NASA, 12, 13, 18, 21, 26, 27, 31, 32 (both), 37, 38
© Scala/Art Resource, 6
© Dr. Seth Shostak/Photo Researchers, 28

For more information, contact
KidHaven Press
27500 Drake Rd.
Farmington Hills, MI 48331-3535
Or you can visit our Internet site at http://www.gale.com

LIBRARY OF CONGRESS CATALOGING-IN-PUBLICATION DATA

Nardo, Don, 1947–
　　Space travel / by Don Nardo.
　　　　p. cm. — (The Kidhaven science library)
　　Summary: Examines the history of people's interest in space travel, the develop-
ment of efforts to leave Earth, and possibilities for colonization and exploration
of outer space.
　　Includes bibliographical references and index.
　　ISBN 0-7377-1406-9 (lib. bdg. : alk. paper)
　　1. Space travel—Juvenile literature. 2. Astronautics—Juvenile literature. 3.
Outer space—Exploration—Juvenile literature. [1. Space travel. 2. Astronautics. 3.
Outer space—Exploration.] I. Title. II. Series.
　　TL793 .N35 2003
　　910' .919—dc21
　　　　　　　　　　　　　　　　　　　　　　　　　　　　　　2002014468

Printed in the United States of America

Contents

Chapter 1

Escaping the Bonds of Earth

In the second half of the twentieth century, humanity began sending various kinds of craft into space. Some carried only cameras, computers, and other instruments. Others had human pilots. Of those carrying people, several landed on the moon and brought back lunar rocks and soil. Others carried the material to build orbiting space stations in which people lived and worked.

Meanwhile, robot craft flew to most of the other planets in our **solar system**. These craft took close-up photos and collected other important data. In the years to come, increasing numbers of both piloted and pilotless spacecraft will continue to travel into space. There is no doubt that humanity stands on the brink of an exciting new era of exploration and knowledge.

The Dream of Flight

Although actual space travel is a fairly recent development, people dreamed about trips beyond Earth for a long time. Indeed, the concepts of flying off into the sky and journeying into space are perhaps as old as humanity itself. Several ancient myths deal with the concept of human flight. Perhaps the most famous is that of a legendary Greek inventor and craftsman named Daedalus. In the story he and his young son, Icarus, are held prisoner on the large Greek island of Crete. To escape, Daedalus constructs wings out of wax and feathers and attaches them to his and Icarus's backs. The two then fly away into the sky. However, Icarus ignores his father's warning not to fly too close to the sun.

An American astronaut explores the moon's surface in one of the NASA missions of the 1970s.

As Daedalus had feared, the wax in Icarus's wings melts and the youth falls to his death in the sea.

Several early writers took the idea of human flight a step further. In the second century A.D., a Greek named Lucian described a great wind carrying some people to the moon. There the travelers find a race of moon people at war with the inhabitants of the sun. They are fighting over who has the right to colonize the planet Jupiter. Another voyage to the moon

An oil painting shows the legendary Daedalus attaching wings to Icarus's back.

appeared in *Somnium,* a 1634 work by German scientist Johannes Kepler. The moon creatures in this story have extremely short life spans. They are born in the morning and die in the evening.

More famous fictional space journeys were those of Frenchman Jules Verne and Englishman H.G. Wells. Verne's *From the Earth to the Moon* was published in 1865 and Wells's *First Men in the Moon* in 1901. In Verne's book, the vessel carrying spacefarers is shot from a giant cannon. In contrast, the spacecraft in Wells's book is coated with a substance that cancels out the effects of gravity. The people in the craft arrive on the moon to find it inhabited by a race of insect-like creatures called Selenites.

Breakthrough at Kitty Hawk

All of these stories were mere flights of fancy, of course. When they appeared, humans had not yet achieved even the simplest kind of powered flight. In fact, most people thought conquering the air was an almost impossible dream. In 1901, the same year Wells published the *First Men in the Moon,* the widely respected *New York Times* predicted that human flight lay 10 million years in the future.

That prediction turned out to be off by 9,999,998 years! A mere two years later, two American inventors, Wilbur and Orville Wright, constructed a primitive airplane. Their great breakthrough occurred on December 17, 1903, at Kitty Hawk, North Carolina.

The Wright brothers make history with their primitive plane at Kitty Hawk in 1903.

Orville managed to fly the plane about 120 feet in twelve seconds.

Gravity and Escape Velocity

Over the years other inventors and pilots flew planes farther, faster, and higher. Yet none of these machines was capable of flying into space. That is because the pull of Earth's **gravity** is very great.

The first scientist to explain how gravity works was Englishman Isaac Newton. In the late 1600s he suggested that all objects exert a force of attraction. He showed in a mathematical formula that this force is directly related to an object's **mass**, the amount of matter it contains. For example, a small object, such as a chair, has very little mass. So it exerts very little gravitational pull and does not attract nearby objects. By contrast, a very large object, such as Earth, has a great amount of mass. So its gravity is capable of attracting chairs, people, mountains, and the moon.

To escape Earth's gravity completely, a craft must reach an extremely high speed. Scientists call the speed at which a body needs to move to escape the gravity of another body its **escape velocity**. Calculations based on Newton's formula show that the escape velocity of Earth is about seven miles per second. That is more than twenty-five thousand miles per hour. Ordinary airplanes cannot travel

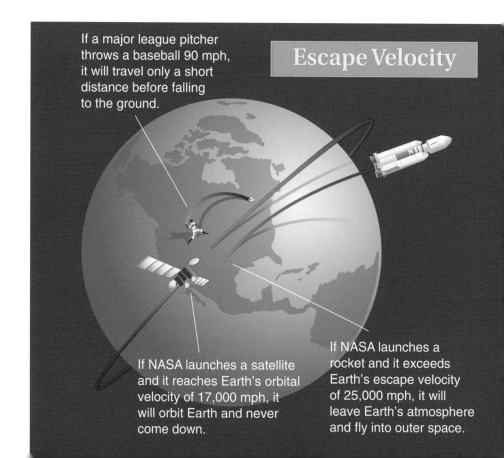

Escape Velocity

If a major league pitcher throws a baseball 90 mph, it will travel only a short distance before falling to the ground.

If NASA launches a satellite and it reaches Earth's orbital velocity of 17,000 mph, it will orbit Earth and never come down.

If NASA launches a rocket and it exceeds Earth's escape velocity of 25,000 mph, it will leave Earth's atmosphere and fly into outer space.

nearly that fast. To fly beyond Earth, therefore, much more powerful machines were needed.

Rockets and the Space Race

The answer turned out to be the rocket engine. The first rockets were invented by the Chinese in the tenth century. Primitive devices that produced a flash of light and loud noise, they were used to frighten enemies in wartime. Later, European and American armies also used rockets in warfare.

Two scientists first proposed using rockets for space travel, a Russian, Konstantine Tsiolkovsky, and an American, Robert Goddard. In 1926 Goddard launched the first rocket powered by liquid fuel. (Before that, rockets had used gunpowder or other solid explosives.) And in 1935 one of Goddard's rockets flew to a height of thirteen thousand feet, more than two and a half miles. Another important rocket pioneer, Germany's Wernher von Braun, moved to the United States in the 1940s. He and several American scientists experimented with powerful rockets in the late 1940s and the 1950s.

Meanwhile, researchers in the Soviet Union were working on large rockets, too. In October 1957 they launched the first artificial satellite into space. Called *Sputnik I*, it orbited Earth once every ninety minutes. Soon afterward, the Soviets launched *Sputnik II*. It carried the first living creature into space, a dog named Laika.

Laika trains in a replica of her capsule shortly before taking her short journey into space.

Not to be outdone, the United States launched its first satellite, *Explorer 1,* in 1958. That same year the nation established the National Aeronautics and Space Administration, or NASA. In the next few years NASA scientists competed with Soviet scientists in the so-called space race. The object was to reach the moon first and dominate the new and "final" frontier of outer space.

Although at first the United States was behind in the race, it soon caught up. In 1961, Alan Shepard became the first American astronaut in space. His flight was part of the Mercury program, which sent astronauts up one at a time. A large rocket carried his **space capsule** high into the sky until it achieved escape velocity. Then the capsule separated from the rocket and flew briefly on its own. Later, the capsule reentered Earth's atmosphere and parachutes slowed its descent until it splashed down into the ocean. There, a ship recovered it and welcomed Shepard aboard.

Many other American space flights followed Shepard's historic trip. Some were part of the Gemini program, in which two astronauts flew together. Finally came the Apollo program. It featured a team of three astronauts in each mission. The high point of both the Apollo program and the space race occurred on July 20, 1969. On that historic day the crew of *Apollo 11* landed safely on the

Technicians prepare Alan Shepard for his sub orbital space flight in 1961.

A mighty Saturn rocket blasts upward from its launch pad during an Apollo mission.

moon, and astronaut Neil Armstrong became the first human being to walk on the moon's surface. The dream of centuries of myths and science fiction had been transformed into reality. At last, the era of space travel had truly begun.

Chapter 2

Colonizing Earth-Space

In a way, the first orbiting satellites and moon landings were the initial baby steps in a huge, almost endless journey. In that epic adventure, humanity will move ever outward into space. Many people will live and work in space. And eventually some will be born and live out most or even all of their lives there.

For some, this concept of humanity as a spacefaring race is difficult to grasp and digest. Over countless generations, people have been born, lived their lives, and died on Earth's surface. And that cycle seems natural—in a way, what was meant to be. The late, great science writer Isaac Asimov coined a term to describe this attitude. He called it "planetary chauvinism," the idea that living beings must originate from and remain on planets.

However, this attitude is slowly but steadily changing. And someday it will seem as quaint as the now discarded idea that people were not meant to fly like birds. Indeed, humanity's conquest of space has already begun. It is a step-by-step process that, naturally enough, began very close to home. The first spacecraft orbited Earth and traveled to the moon, which is very near compared to the other planets. Scientists commonly call the general vicinity of the Earth and moon "Earth-space." It is in this zone that the first space industries are developing, and it will be the location of the first human space settlements.

American astronaut Buzz Aldrin stands on the moon, the first major hurdle in colonizing space.

Space Labs and Stations

The **industrialization** (creation of industries) and settling of Earth-space began in a small way in the 1970s. The Soviets put the first space lab—Salyut I—into orbit in 1971. And a seventy-ton American version—Skylab—began operation in 1973. The Skylab astronauts orbited Earth at an altitude of 270 miles. They conducted studies of the sun and Earth's weather patterns. They also studied the effects of weightless conditions on the human body so that future space workers would know what to expect.

Similar experiments occurred on U.S. space shuttle missions. NASA launched the first shuttle in 1981. The shuttles were designed to be "space trucks"—to carry satellites and other heavy objects into or back from space.

The next major step was the Russian space station Mir in 1986. Mir remained in operation for fifteen years before it became outdated. In that time it orbited Earth 86,325 times. That gave scientists the opportunity to study weightless conditions over long periods. Russian astronaut Sergei Avdeyev still holds the record—a total of 747 days in orbit over the course of three missions. In addition, seven American astronauts spent time aboard Mir. They helped with some of the twenty three-thousand scientific experiments the station conducted.

The lessons of these missions and experiments made an even larger space venture possible. In

1998 the first **module** of the International Space Station, or ISS, was launched. Each year several new modules are added, piecing the station together like a huge Tinkertoy set floating 250 miles above Earth. The ISS is scheduled for completion in 2006. It is sponsored by sixteen nations—the United States, Russia, Canada, Japan, Brazil, and eleven European countries.

Science and industry will thrive in the ISS's labs. Astronauts will grow crystals and other materials in weightless conditions. This will allow the development of new kinds of medicines to fight disease. Experiments with flames and melting metals in

Aboard Skylab in 1973, an astronaut monitors his companion for the effects of weightlessness.

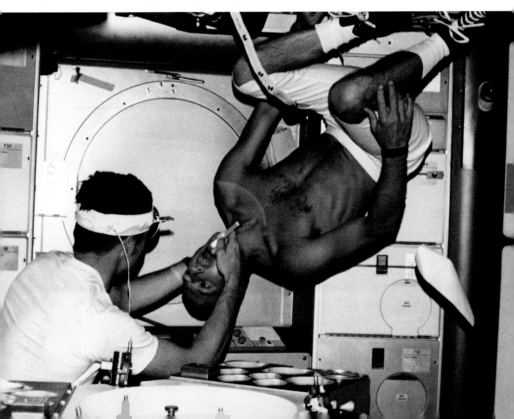

The crew of the International Space Station poses while floating weightless.

space will lead to better computer chips and other sorts of mechanical parts, and new, cleaner forms of energy will be explored and perfected.

Colonies in Orbit

Other stations like the ISS will be constructed in Earth-space. Over time, some of these will grow into actual cities, often called space colonies. These will be designed for more than just temporary visits by teams of scientists, as in the case of the ISS. In addition to labs and workshops, there will be apartments, shops, restaurants, parks, movie theaters, and small patches of farmland. At first, only a few hundred people will live in such colonies, but that number will grow. Eventually some space colonies may support tens of thousands of people on a permanent basis.

Unlike the astronauts on Skylab and the space shuttles, the inhabitants of a typical space colony will not be weightless. Instead, they will have gravity that feels almost identical to that of Earth. The artificial gravity will be created by spinning the colony, producing an effect called **centrifugal force**. An earthly example would be someone swinging a water-filled bucket in circles. Held in place by the force of the swing, the water will not spill out, even when the bucket turns upside down. Similarly, people living in the outer sections of a spinning space

An artist depicts a large space station floating in Earth-space.

colony will feel a gravity-like effect. Moreover, centrifugal force moves outward from the center. Therefore, objects located in the center of the city will be weightless. That is where zero-gravity experiments and manufacturing will take place.

As for where such colonies will be located in Earth-space, most experts say the Lagrangian points would be ideal. They are named after French astronomer Joseph-Louis Lagrange. In 1772 he used mathematics to show that five special points exist within the interacting gravity fields of Earth and the moon. Objects orbiting at these points keep in perfect step with these bodies. Also, objects at points L4 and L5 remain "locked" in place. People living on a colony at L4 would always see the same sides of Earth and the moon. This stable area would make it easier for spacecraft to locate and land at the space colony. So ferrying people and supplies to colonies located at L4 and L5 would be relatively simple and cheap.

Lunar Mining Bases

It is likely that the moon will also be colonized. At least for the foreseeable future, however, lunar colonies will likely be small. The moon lacks nitrogen and some other important elements necessary for sustaining life. It does seem to have tiny amounts of water trapped beneath the rocks in its polar regions. But there is probably not enough to

Scientists think a mining operation on the moon will look something like this.

meet the needs of entire cities for very long. That means that certain vital materials will have to be flown in from Earth, which will be expensive.

So at first moon colonies will likely consist of small mining bases. The moon does have large amounts of soil and minerals that could be used in constructing orbiting space cities. It would be much easier and cheaper to get these raw materials from the moon than from Earth because the moon's gravity is much weaker than Earth's. Experts estimate it takes only one-twentieth the energy to lift materials off the moon than off Earth.

Furthermore, expensive rockets may not be necessary for this task. Plans have already been created to build **mass drivers** on the moon. A mass driver consists of a long track or tube made of aluminum or some other metal. A bucket containing the material one wants to launch is placed in the track. Then magnets or an electric current cause the bucket to **accelerate**, or move faster and faster, down the track. When the bucket reaches the end, it flies off into orbit around the moon. There, a space shuttle or other craft can scoop it up and ferry it to its intended destination.

Eventually, moon bases and orbiting colonies will do more than serve the needs of people living in Earth-space. They will also become the jumping-off points for the next stage of space travel—exploring the rest of the solar system.

Chapter 3

Exploring the Solar System

The ongoing adventure of colonizing Earth-space will continue for several generations. Yet even more exotic and exciting aspects of space travel await humanity. Beyond the Earth-moon system lie the many other objects that orbit our star, the sun. These include the other planets, their moons, and the **asteroids** and comets. Exploration of these wonders has already begun, and in the next few decades and centuries, the entire solar system will be charted and studied in detail. In addition, humans will harvest and utilize some of that system's vast array of natural resources.

Missions to the Planets

The first attempts to study the other members of the sun's family began in 1962. Between that year and 1975, NASA launched several spacecraft in the Mariner series. The Mariner probes were launched into space by rockets and carried equipment that allowed scientists to communicate with them using

The Solar System

Venus

Comet

Sun

Mercury

Jupiter

Earth

Mars

Asteroid Belt

Saturn

Uranus

Neptune

Pluto

radio signals. The probes were designed to fly by various planets and snap close-up photos. *Mariner 2 and 5* successfully reached Venus. Venus is the second planet from the sun, Mercury being the first and Earth the third. Another Mariner probe photographed both Venus and Mercury. *Mariner 4, 6,* and *7* flew by Mars, the so-called red planet, which is the fourth planet from the sun. *Mariner 9* actually went into orbit around Mars. The photos it sent back to Earth showed many large craters and several tall volcanoes.

Clay, Storms, and Moons

Humans have always been fascinated by Mars, and many writers and scientists have wondered if there might be life there. So that planet was the object of another series of spacecraft—the Vikings. *Viking 1* landed on Mars in July 1976 and *Viking 2* followed two months later. These probes took stunning photos of the Martian surface. They also collected soil samples, analyzed them in small portable onboard labs, and sent the results via radio signals back to Earth. They found the soil in the landing areas to be an iron-rich clay. The Viking craft also measured Mars's atmosphere and observed Martian storms. No evidence of life was found. (But a small chance remains that some kind of life might exist somewhere else on the planet.)

Shortly after the Vikings touched down on Mars, NASA launched *Voyager 1* and *2*. Their mission was to study the planets lying beyond Mars—Jupiter, Saturn, Uranus, and Neptune. Astronomers call them the gas giants partly because they are very large and also because they are made up largely of thick gases and liquids. *Voyager 1* reached Jupiter in March 1979. *Voyager 2* arrived in July. Both probes collected data and then moved on to Saturn. *Voyager 2* went even further, reaching Uranus in 1986 and Neptune in 1989. These craft achieved spectacular success. They discovered twenty-two planetary moons, found faint rings around Jupiter,

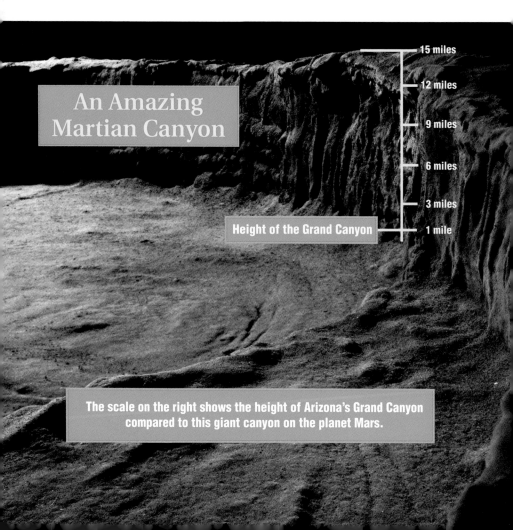

An Amazing Martian Canyon

15 miles
12 miles
9 miles
6 miles
3 miles
Height of the Grand Canyon
1 mile

The scale on the right shows the height of Arizona's Grand Canyon compared to this giant canyon on the planet Mars.

The asteroid Ida and its tiny moon, Dactyl, as photo-graphed in a close flyby in 1993.

watched volcanoes erupt on Jupiter's moon Io, and photographed giant storms on Neptune.

Mining the Asteroids

In 1995 still another space probe, dubbed *Galileo*, studied Jupiter up close. *Galileo* also flew very near to two asteroids, Gaspra and Ida. Asteroids are hunks of rock and metal ranging in size from large boulders to mountains. A few are even bigger. The largest known asteroid, Ceres, is 567 miles (914 km) in diameter, about the size of the state of Texas. The vast majority of these bodies lie in the region between the orbits of Mars and Jupiter, which astronomers call the asteroid belt.

In an artist's view, an asteroid has been towed into Earth's orbit to be processed for its metals.

The asteroids have tremendous potential value to humans. Because these objects are rich in various metals, mining them would conserve supplies on Earth, which are rapidly dwindling. Also, processing metals (separating them from rocks and dirt) pollutes Earth's air, soil, and water. In the case of asteroid mining, this problem would be eliminated because the processing would take place in space.

Another benefit of mining asteroids is that it would require little energy. Most asteroids are small and have extremely tiny gravities. A mining ship would not have to waste precious fuel landing. Instead, it would park beside the asteroid. Also, the miners and metals would be nearly weightless, so getting them off the asteroid would require little or no fuel.

A number of scientists have estimated what such a mining operation might be worth. An asteroid just one hundred yards wide—the length of a football field—contains about 3 million tons of material. An operation lasting only a few weeks could collect up to ten thousand tons of material. Now consider an asteroid six-tenths of a mile (one kilometer) across. Its supplies of nickel and iron alone would be worth at least $500 billion (enough to pay each man, woman, and child in the United States $178,000). It would also yield precious metals such as platinum and gold. In addition, it would contain large amounts of sulfur, aluminum oxide, and other useful minerals. Together, the materials mined from this single small asteroid would be worth several trillion dollars. Eventually, space colonies, satellites, and even structures on Earth itself will be constructed from materials gathered from asteroids.

Searching for Life

While crews of miners tap the resources of the asteroid belt, other space travelers will continue to explore the solar system's many wonders. Some missions will venture out beyond the gas giants. Spacecraft from Earth will finally reach the ninth and outermost planet—Pluto. Orbiting the sun more than thirty-nine times farther out than Earth, Pluto is a tiny world only about two-thirds the size of Earth's moon. A robot craft that lands on Pluto may confirm what astronomers suspect—that the planet is so cold its atmosphere lies frozen on its surface.

Pluto is almost surely too cold to support any sort of life. But humans will still continue searching for life in other parts of the solar system. Mars remains a possibility. Perhaps bacteria or other primitive forms of life exist deep underground. New missions to Mars are already being planned. The Jet Propulsion Lab in Pasadena, California, has already designed special robots to dig down into the Martian soil. Besides looking for life, these mini-bulldozers will help excavate foundations for buildings that humans will need. For there is little doubt that people will eventually colonize the red planet, because it is relatively close, relatively warm, and has water and an atmosphere.

The Liquid Ocean

Even more advanced forms of life might exist in places in the solar system that contain large amounts of water. Promising in this respect is Europa, one of the four largest moons of Jupiter. Most astronomers

This device, known as the Mars Rover, will travel over the planet's surface to search for natural resources and life.

and NASA scientists believe that Europa has a liquid ocean beneath its icy outer shell. The moon's core is warm, so conditions there might be right for some kind of aquatic (water-based) life. An ocean of relatively warm water may also lie beneath the icy shell of Neptune's largest moon—Triton.

To explore these mysterious inner seas will require special devices. Scientists at the Jet Propulsion Lab are presently building what they call "cryobots." Each is about three feet long and looks something like a torpedo. A spacecraft will drop a cryobot onto Europa's surface. Warm water in the device's nose will begin melting the ice, allowing the

This drawing shows how a cryobot (inset) will penetrate Europa's icy surface (top) and send out a detection device (center).

probe to sink deeper and deeper. When the cryobot reaches the inner ocean, onboard lights, cameras, and detectors will search for life.

It is possible, of course, that humans may find that no other life exists in the solar system. But if so, the search for life beyond Earth will not end. There is little doubt that someday humanity will bravely begin the next phase of space exploration—journeys to other solar systems.

Chapter 4

Voyages to the Stars

Almost everyone is familiar with the **interstellar** journeys shown in the *Star Trek* and *Star Wars* films. (The term *stellar* refers to stars, so interstellar means moving from one star to another.) The space travelers in these films find many planets moving around distant stars. They also encounter life of various kinds on these planets. Until recently, such adventures were seen as pure fiction. Most people, including most scientists, did not think they could ever happen.

However, that view is rapidly changing. In the 1990s astronomers first found planets revolving around other stars. By early 2002 about eighty such planets had been found, and the total continues to grow. Scientists now believe that many, if not most, stars in the universe have at least some planets.

Of course, at present no one knows if life exists on any of these alien planets. But the human thirst for knowledge and spirit of exploration is strong. And it is probably only a matter of time before Earth sends spacecraft on voyages to the stars. Such craft will

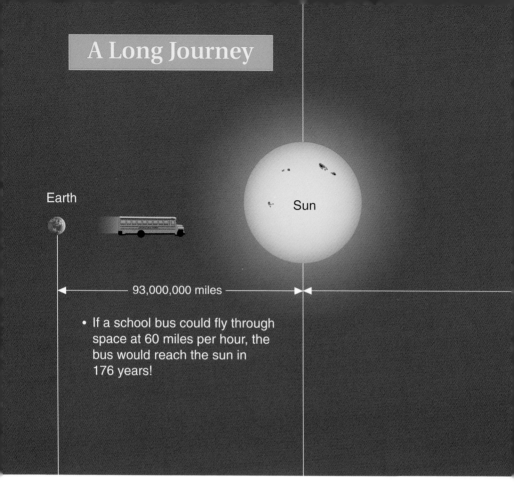

Earth

Sun

← 93,000,000 miles →

- If a school bus could fly through space at 60 miles per hour, the bus would reach the sun in 176 years!

not only search for life, but also begin to colonize other solar systems.

Incredible Distances

Before such exciting journeys can begin, however, some difficult problems must be overcome. The first consists of the incredible distances that must be covered. They are so huge that trying to express them in standard units of measure such as miles or kilometers is confusing and meaningless. Astronomers use a special and much larger unit called a **light year**. This is the

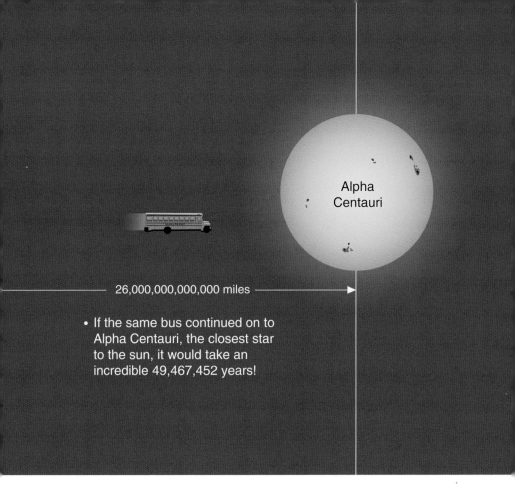

Alpha
Centauri

←————— 26,000,000,000,000 miles —————→

- If the same bus continued on to
  Alpha Centauri, the closest star
  to the sun, it would take an
  incredible 49,467,452 years!

distance that light, which moves at 186,000 miles per second, travels in a year—about 6 trillion miles.

The nearest star to the sun, Alpha Centauri, is 4.3 light years, or about 26 trillion miles, away. That means that when we look up at Alpha Centauri tonight, the light we see left it 4.3 years ago. If that star somehow disappeared tonight, people on Earth would not know about it for more than four years. During that time they would still be seeing the light that left the star *before* it disappeared!

Now consider space travelers from Earth attempting to reach Alpha Centauri. Their ships will not be able to attain even a significant fraction of the speed of light. The Pioneer and Voyager probes have now left the solar system and are moving away at about twenty miles per second. That is very fast by earthly standards, but at that speed it would take space travelers some forty thousand years to reach Alpha Centauri.

Also one must keep in mind that Alpha Centauri is the *nearest* star to our sun, only two among hundreds of billions of stars in our **galaxy**. A galaxy is a gigantic group of stars all held together by gravity. Astronomers call our home galaxy the Milky Way. It is a big, more or less circular disk about one hundred thousand light years across, so a spacecraft traveling at twenty miles per second would require a billion years to make it across the Milky Way.

Problems with Speed, Time, and Sleep

Clearly, such long journeys would be impractical and not worth the effort. The most obvious way to overcome the problem would be to find a way of making spacecraft move faster. This is often the solution in the world of *Star Trek* and other fictional realms. The starships are equipped with "warp drive," "hyperdrive," or similar devices that allow them to travel faster than light.

However, most scientists believe that in the real world nothing can travel faster than light. And even if humans *could* travel that fast, they would have to deal with extremely odd changes in time. On their rapidly moving ship, time would seem to pass normally. But in the outside world millions of years would have gone by. When the travelers return to Earth all the people they had known would be long dead. They might even find the entire human race gone.

The Milky Way likely looks very much like this distant galaxy—a huge pinwheel containing billions of stars.

A number of scientists and writers have thought about other ways to make interstellar travel practical. One possibility is to have the space travelers **hibernate**, or go to sleep for long periods. They could go to sleep shortly after leaving our solar system and wake up shortly before entering the target system. To make this work, the computers and other machines on the ship will have to be very advanced. They will have to run the ship, keep it on course, and

Advanced computers aboard an interstellar space ship monitor the craft's systems while the crew sleeps.

repair it if it is damaged. They will also have to monitor carefully the health of the sleeping crew.

Even if such space hibernation is perfected, it will be practical only for trips to the nearest stars. Suppose that the crew of a starship wants to travel to a star a thousand light years away (only one-hundredth the width of the galaxy). Also suppose that the ship can attain a speed of twenty thousand miles per second, a thousand times faster than that of the Voyager probes. The journey would still take about ten thousand years. Even if people's life spans could be greatly extended, it may not be either possible or desirable for people to sleep that long.

Colonizing the Galaxy?

A more realistic approach to interstellar travel would be a **generation ship**. This would be a large space vessel or colony in which many generations of people would live and die during the journey. Returning to Earth would not be part of the original plan. Instead, the goal would be for the great-great-great-grandchildren of the first crew to reach another solar system and there establish one or more human colonies.

To support hundreds or thousands of people for generations would require such a ship to be a self-contained little world. Ideally, it would be at least several miles long. It would have farmland, lakes, towns, and even its own weather. A number of

An artist's view of the inside of a gigantic generation ship shaped like a cylinder.

shapes for the vessel would be possible, but perhaps the best would be a cylinder, like a soup can. The people would live on the curved inside surface. They and their animals and houses would stay on the "ground" thanks to artificial gravity created by spinning the ship.

When the ship finally reaches the new star, the inhabitants might search for a suitable planet to colonize. Or they may feel more comfortable building and living in space cities instead. After all, none of them would know what it is like to live on a real planetary surface. After a while, some of the colonists might build new generation ships. These would strike out for other stars. And the process might keep repeating itself until humans had colonized the whole galaxy. This would likely take millions of years.

Epic Journeys

This epic, romantic journey of humanity may never come to pass, of course. The first interstellar ships and colonies may be wiped out by disease, wars, or other disasters. Or humanity may well find out it is not the only intelligent race in the galaxy. And the inhabitants of other solar systems may see human colonists as intruders. Another possibility is that an alien race is itself already in the process of colonizing the galaxy. Perhaps someday two great ships, one human, the other alien, will meet in the dark gulfs of interstellar space. If so, the hope is that the high intelligence that allowed each to master space travel will foster a spirit of peace and cooperation.

accelerate: To move faster and faster.

asteroids: Small stony or metallic bodies orbiting the sun, most often in the asteroid belt (lying between Mars and Jupiter).

centrifugal force: The outward-moving force created when an object is swung or spun in circles.

escape velocity: The speed at which a body needs to move to escape the gravity of another body.

galaxy: A gigantic group of stars held together by their mutual gravities. Our galaxy is called the Milky Way.

generation ship: A huge vessel or colony on which several generations of space travelers will live and die during a long journey.

gravity: A force exerted by an object that attracts other objects. The pull of Earth's gravity keeps rocks, people, and houses from floating away into space. It also holds the moon in its orbit around Earth.

hibernate: To sleep for a long time.

industrialization: The creation of industries.

interstellar: Between stars, usually referring to travel from one star to another.

light year: The distance that light travels in a year, or about 6 trillion miles.

mass: The measurable amount of matter making up an object.

mass driver: A device that uses electricity and/or magnetism to hurl objects down a long track or tube and launch them into space.

module: Individual piece of a larger structure.

solar system: The sun and all the planets, moons, asteroids, and other objects held by the sun's gravity.

space capsule: A small container in which astronauts ride. Usually the capsule is carried into space by a rocket, then separates from the rocket and flies on its own.

stellar: Having to do with stars.

For Further Exploration

Pam Beasant, *1000 Facts About Space*. New York: Kingfisher, 1992. An informative collection of basic facts about the stars, planets, asteroids, and other heavenly bodies.

Joanna Cole, *The Magic School Bus: Lost in the Solar System*. New York: Scholastic, 1990. Grade schoolers will enjoy this story of a teacher who takes her students on a journey to the planets in a magic bus.

Kenneth C. Davis, *I Don't Know Much About the Solar System*. New York: HarperCollins, 2001. An exciting, fact-filled tour of the solar system, set in a question-and-answer format.

Nigel Henbest, *DK Space Encyclopedia*. London: Dorling Kindersley, 1999. This beautifully mounted and critically acclaimed book is the best general source available for grade school readers about the wonders of space.

Robin Kerrod, *The Children's Space Atlas: A Voyage of Discovery for Young Astronauts*. Brookfield, CT: Millbrook, 1992. A well-written, informative explanation of the stars, planets, comets, asteroids, and other objects making up the universe.

Don Nardo, *The Solar System*. San Diego: KidHaven, 2002. This colorfully illustrated book describes the various members of the sun's family and ends by examining the probability that the solar system will eventually be absorbed by the giant black hole lurking in the center of our galaxy.

Sally Ride and Tam O'Shaughnessy, *The Mystery of Mars*. New York: Crown, 1999. A commendable exploration of the red planet, the book compares Mars to Earth, discusses the possibility of life on Mars, and tells about the various space missions that have explored that planet.

Kathy Ross, *Crafts for Kids Who Are Wild About Outer Space*. Brookfield, CT: Millbrook, 1997. Aimed at grade schoolers, this book provides them with hands-on activities that are also educational.

Deborah A. Shearer, *Walking on the Moon*. Mankato, MN: Bridgestone, 2002. Tells how human beings built and piloted spacecraft to Earth's nearest neighbor in space.

Gregory L. Vogt, *Asteroids, Comets, and Meteors*. Brookfield, CT: Millbrook, 1996. Tells the basic facts about these stony, metallic, and/or icy bodies orbiting the sun.

Index

About the Author

In addition to his acclaimed volumes on ancient civilizations, historian Don Nardo has published several books about the wonders of space. Among these are studies of the asteroids, the moon, Neptune, Pluto, the solar system, and black holes. Mr. Nardo lives with his wife, Christine, in Massachusetts.